Graham King has been an experienced Surrey Area of NAFAS tutor and demonstrator for almost thirty years He is a fully qualified and experienced florist with over twenty years of experience. He teaches and demonstrates using a large variety of creative floral designs and crafts passing on fascinating ideas and techniques to students and audiences alike. He is also a skilled self-taught arts and crafts expert with a vast knowledge of many creative craft subjects. One of his passions, favourite subjects and specialities is experimental textile work, where he makes fantasy flowers, foliage and containers from varied materials.

This book is for Ilona Barney. A great friend and a mentor sadly missed. Ilona was a brilliant NAFAS demonstrator and competitor who encouraged me to write this book. She taught me many floral design skills, especially to be brave and experiment with colour.

Graham King and Stephen Barney

EVERY FLOWER HAS ITS PLACE

AUSTIN MACAULEY PUBLISHERS™

LONDON * CAMBRIDGE * NEW YORK * SHARJAH

Copyright © Graham King and Stephen Barney 2024

The rights of Graham King and Stephen Barney to be identified as authors of this work has been asserted by the authors in accordance with sections 77 and 78 of the Copyright, Designs and Patents Act 1988.

All rights reserved. No part of this publication may be reproduced, stored in a retrieval system, or transmitted in any form or by any means, electronic, mechanical, photocopying, recording, or otherwise, without the prior permission of the publishers.

Any person who commits any unauthorised act in relation to this publication may be liable to criminal prosecution and civil claims for damages.

The story, experiences, and words are the author's alone.

A CIP catalogue record for this title is available from the British Library.

ISBN 9781398486676 (Paperback)
ISBN 9781398486683 (Hardback)
ISBN 9781398486690 (ePub e-book)

www.austinmacauley.co.uk

First Published 2024
Austin Macauley Publishers Ltd®
1 Canada Square
Canary Wharf
London
E14 5AA

I would like to start by acknowledging the tremendous effort and support of Stephen Barney my co-author, for all his brilliant photography work on this project and the many hours put in to setting up each shoot, and in editing every image. For without Stephen's artistic eye and creative talent in photography, this book would never have been possible, thank you!

Next, I would like to say a massive thank you to Claire Barney, for putting her many literary talents to good use with this book, by helping me edit, proofread, and re-write every draft. Her unique insights into the world of publishing and creative writing have helped steer this book to completion and have been invaluable throughout this whole process.

Another, big thank you to Frances Collins for being a fantastic sounding board when drafting parts of this book and letting me use your garden and out-buildings as perfect locations for my arrangements.

Also, a big thank you to Dr Christina Curtis, whose botanical knowledge has been an enormous help to me throughout the making of this book, and for checking my botanical writing during the editing process; your help has been greatly appreciated.

And lastly, a huge thank you to all the 'Behind the scenes' dedicated helpers who have given their time and effort to help fetch and carry, help prepare the work, offer me a location and for braving the elements, especially the wind, rain and mosquitoes, alongside me.

Vic and Wendy King, Terry and Nick Cornell, Sally Lee, and Dorota Chalupczak, Wendy and Claire Barney, Bill and Judy Bolt, and Linda Dowding.

All the staff and volunteers at the Whitchurch Silk Mill, Hampshire. whitchurchsilkmill.org.uk

John Cooksley (Blacksmith) Rural Life Centre, Tilford, Surrey. rural-life.org.uk

Mr and Mrs R. Thompson.

Great Fosters Hotel, Egham, Surrey. greatfosters.co.uk

Mid Hants Railway Watercress Line. watercressline.co.uk

St Nicholas Church, Compton, Surrey.

St Mary's and All Saints Church, Dunsfold, Surrey. dunsfoldchurch.co.uk

All Saints Church, Witley, Surrey allsaintswitley.org.uk

Table of Contents

Foreword	15
Introduction	16
Chapter 1: Flowers of Industry	17
The Mill Race	19
Materials Used:	20
The Water Wheel	21
Materials Used:	22
The Wheels of Industry	23
Materials Used:	24
On the Loom	25
Materials Used:	26
With These Threads	27
Materials Used:	28
Basket of Silks	29
Materials Used:	30
Of Silk and Wool	31
Materials Used:	32
The Weaver's Window	33
Materials Used:	34
The Carpenter's Workshop	35
Materials Used:	36
In the Forge	37

Materials Used: 38

The Blacksmith's Workshop 39

Materials Used: 40

Forged Flowers 41

Materials Used: 42

Chapter 2: In My Secret Garden 43

Gardener's Delight 45

Materials Used: 46

A Garden Riddle 47

Materials Used: 48

Cascading Cans 49

Materials Used: 50

Vision in Blue 51

Materials Used: 52

The Garden Room 53

Materials Used: 54

Fresh for the Table 55

Materials Used: 56

The Water Sculpture 57

Materials Used: 58

The Potting Shed 59

Materials Used: 60

The Flower Clock 61

Materials Used: 62

A Floral Candelabrum 63

Materials Used: 64

High Tea 65

Materials Used:	*66*
Mid-Summer Pedestal	**67**
Materials Used:	*68*
Chapter 3: Coastal Inspirations	**69**
The Seahorse	**71**
Materials Used:	*72*
Sails	**73**
Materials Used:	*74*
The Beachcomber's Bounty	**75**
Materials Used:	*76*
Caught in the Breakwater	**77**
Materials Used:	*79*
Under the Sea	**80**
Materials Used:	*81*
Flotsam and Jetsam	**82**
Materials Used:	*83*
The Spider Crab	**84**
Materials Used:	*85*
The Mermaid's Pearls	**86**
Materials Used:	*87*
Evening Glow	**88**
Materials Used:	*89*
The Beach Hut	**90**
Materials Used:	*91*
The Lobster Pot	**92**
Materials Used:	*93*
A Seaside Treat	**94**

Materials Used:	*96*
Chapter 4: Woodland Wonderland	**97**
The Magenta Tower	**99**
Materials Used:	*100*
The Stag	**101**
Materials Used:	*102*
The Old Gate Post	**103**
Materials Used:	*105*
The Enchanted Tree	**106**
Materials Used:	*107*
The Fairy Ring	**108**
Materials Used:	*109*
The Woodland Bridge	**110**
Materials Used:	*111*
Fallen	**112**
Materials Used:	*113*
Butterflies	**114**
Materials Used:	*115*
Mother Nature's Crown	**116**
Materials Used:	*117*
Mushrooms and Toadstools	**118**
Materials Used:	*119*
The Woodcutter's Cache	**120**
Materials Used:	*121*
Beauty on the Brook	**122**
Materials Used:	*123*
Chapter 5: Times Gone By	**125**

The Master's Study   **127**
   Materials Used: *128*

The Fallen Abbey   **129**
   Materials Used: *130*

Call the Porter   **131**
   Materials Used: *132*

A Christmas Wedding   **133**
   Materials Used: *134*

A Floral Tribute   **135**
   Materials Used: *136*

Floral Teacup   **137**
   Materials Used: *138*

Homage to Grinling Gibbons   **139**
   Materials Used: *140*

Harvest Time   **141**
   Materials Used: *142*

A Country Affair   **143**
   Materials Used: *144*

Antique Fabric   **145**
   Materials Used: *146*

Miss Havisham's Wedding   **147**
   Materials Used: *149*

Left Luggage   **150**
   Materials Used: *151*

A Final Word   **152**

Foreword

Flower arranging is a very enjoyable hobby. Through it, we can release our inner creativity and enhance our wellbeing. It gives me great pleasure to be asked to write regarding 'Every Flower Has Its Place'. The author, Graham King, has been a demonstrator and teacher in the Educational Charity NAFAS (National Association of Flower Arrangement Societies) for more than 25 years. Over that time, I have seen his floral artwork develop and his deep passion for craft evolve.

Graham is also a keen competitor in floral art competitions, winning many awards and has also produced exhibition pieces. His designs now always incorporate beautiful mixed media craft items along with floral materials giving his designs unique styling and detailed embellishments.

In this book, Graham has staged his designs in a variety of locations from the beach, gardens, and woods to the more unusual sites such as on a waterwheel and a wall in a ruined abbey. The descriptions of his designs give valuable insight into Graham's world of creativity with techniques and tips to inspire others.

Graham's wonderful creations have been skilfully captured by professional photographer Stephen Barney, son of Ilona and Maurice Barney, whose realistic hand-crafted balsa wood 'Barney Birds' are now treasured collectables especially in many flower arrangers' homes.

I know this book, an exciting visual resource of ideas, will be inspiration for many encouraging readers to 'have a go' themselves.

Dr Christina D. Curtis
NAFAS National Demonstrator and Judge
Vice President of the Surrey Area of NAFAS
April 2021

Introduction

Welcome to my colourful world of excitement through floral arranging and craft work which have much inspired me to write this book. It has been a fantastic journey of discovery, hard work and an education for me.

The seed was planted when I was showing a photo of a design that I had made to a colleague who suggested that I should write a book to pass my excitement and enthusiasm on to you, so here is the result.

I have been crafty from a very young age and enjoyed painting and making things. I have also always enjoyed wandering around gardens, studying nature and plants. I never realised that one day I would end up as a florist, flower arranger, tutor and demonstrator. I am a very active member of NAFAS (The National Association of Flower Arrangement Societies) and help as much as I can within the Surrey area where I have been an Area Demonstrator now for 25 years.

I very much hope that you enjoy my new book as you turn every page and maybe I might inspire you also to arrange some flowers yourself. Go on, have a go and have as much fun as I have had and still do!

Happy arranging!

Chapter 1
Flowers of Industry

The Mill Race

When I first visited the Silk Mill, I carefully had a good look around and took lots of pictures, notes, and measurements. As I saw the mill race, I could see in my imagination something growing out of the mossed bank, something that would go with the surroundings but also something that would stand out! So here I thought that this would be a good spot for some homemade flowers and leaves.

I started with some ivy trails, made from a roll of matted together sisal, painted to imitate variegated ivy and these were glued onto a wire stem. The larger leaves were made from a sandwich of the same sisal material and tissue paper glued on with PVA glue, then also painted with artists' acrylic paints. I prefer acrylic paints as they are water resistant when completely dry.

The strange tropical flowers are made from a wire mesh petal shape covered with red velvet material on the inside and dried moss glued onto the outside. (I cannot live without my glue gun!) The stamens of these strange blooms are made from segments of dried cotton flower seed heads, painted, and glued onto embroidery thread-wrapped wires. Once I had made all the components of this design, I arranged them in a block of dry floral foam which was in turn securely taped and glued to a heavy wooden board, to weigh it down and prevent it from falling in the fast-flowing River Test. (Although I was extremely careful with all my tools and equipment, I did manage to lose my favourite wire cutters in the river!)

Materials Used:

Dried carpet moss, dried lotus seed heads, sisal string, pressed sisal abaca wrap, cotton seed head segments, PVA glue, artists' acrylic paint, and assorted stub wires.

The Water Wheel

While on a previous visit to the fantastic Silk Mill at Whitchurch, we were chatting to the museum manager and jokingly mentioned it would be great to decorate the waterwheel. The manager thought it would be a good idea as nobody has ever decorated the wheel before with flowers. So, we booked a date, and took some measurements and a few photos to draw up a plan.

We decided to arrange the flowers in a tumbling cascade style which resembles the motion of the wheel with water splashing everywhere. The wooden slats of this wheel are just like a series of L-shapes like shelves on a bookcase, ideal to hold trays of soaked floral foam. The wheel is very large, and we could only just reach the top to arrange the flowers. I am glad there were two of us, one to climb the wheel, and the other to pass up the flowers and foliage.

We chose some bright colours that would contrast and complement such an old historical building. We also chose lots of light fluffy, delicate textures to resemble water droplets. To give movement to the design we used lots of *Asparagus* fern and included torn lengths of bleached mulberry bark.

We had great fun creating this design although it was really hard work. It was also a very tight space to work in, and luckily, no one fell in the water!

Materials Used:

Peach parrot tulips, white *Astrantia, Asparagus* fern, bleached *Morus* (mulberry) bark, *Gypsophila paniculata* 'Million Stars', Singapore orchids, pink *Hydrangea*.

The Wheels of Industry

To create a feeling of movement and motion a curved or S-shaped design best portrays this. I wanted to give the impression of cogs and wheels whirring in motion with flowers wrapped around a cog and wheel.

The flower arrangement is made from polystyrene spheres which I glued in a continuous spiral of string. You will need to test glues on polystyrene first. Some glues, especially solvent based will melt and eat it away. A 'Tacky' PVA or cool melt electric glue gun would be suitable. The roses are made from a sinamay ribbon held together with wires. The centre of the swirl design is arranged in OASIS® dry foam block, and the rest are simply glued in place directly onto the main spindle with the glue gun. To add to the movement and help to balance the visual weight of the design, I added two dried New Zealand Hapene leaves which are also held in place with the glue gun.

To create this industrial-themed background, I re-used an old painted canvas background. I cut out oblong-shaped sheets of painted canvas glued together. The rivets are made from stick-on imitation plastic pearls and painted over with acrylic paints. To give the effect of water running down, I poured gloss varnish down the canvas.

The pipe work and cogs are made from plastic pipes and wooden cogs painted with a mixture of acrylic paint and silver sand to give a rusty metal effect.

Materials Used:

Dried Hapene (*Phormium tenax*) leaves, natural sinamay ribbon, jute and hemp string.

On the Loom

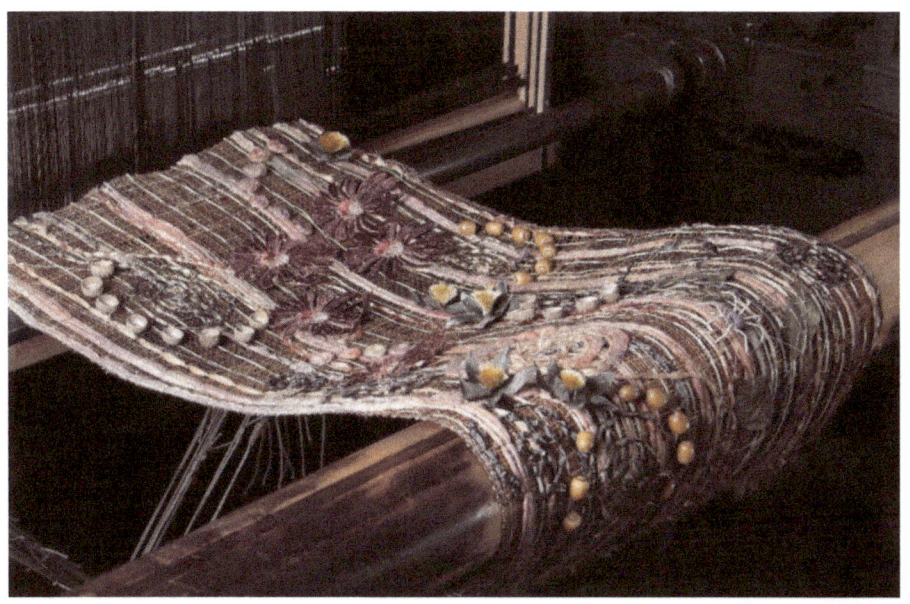

After seeing the weaving room at the Silk Mill, I was totally inspired, and many ideas filled my head. One idea was to make a design that looks as if it was being woven on a loom and so I set about and made this wall hanging using different techniques. The influence for colours came from the hand-dyed wool that I found which I hand stitched in lengths onto the background of woven plant fibres.

I used lines and spirals to give movement to the design, otherwise, it could appear static and I have also included some homemade silk flowers and silver spider's webs. (Can you see the crafted spider?) I then stitched on a few strips of silk fabric which adds a variation of texture and, to give another dimension and contrast in shape, I glued on some acorn cups and acorns.

The larger fabric flowers towards the top of the design are cut from stiffened machine embroidered material. I really enjoyed producing this design as craftwork has always been my favourite subject and I love making objects from a limited range of materials.

This was intentional as it really makes you work hard at those thought processes and come up with imaginative ways of using your chosen materials. It is healthy for you to expand your powers of thought. (As an old friend says, 'Use it or lose it!')

Materials Used:

Natural plant fibre mesh, acorn cups and acorns, selection of lengths of silk fabric, coloured wool, embroidery thread, silver thread, and cut-out fabric flowers.

With These Threads

It is amazing to think that such a beautiful gossamer thread is spun from a silkworm that eats mulberry leaves. Pure silk fabric is a fantastic medium to work in as you can make it into so many different things. It is available in some of the most beautiful colours in the world and is woven in a variety of weaves that produce wonderful textures. My favourite silk fabric is Dupion which has a little added texture and makes great silk flowers.

Here I have made a small corsage-style spray of silk flowers which is gently resting on a row of wooden bobbins holding silk thread. I have made a few silk leaves on which are embroidered veins in chain stitch. These have been highlighted with a little watered-down artist's acrylic paint and then wired with a fine rose wire before being finely bound with embroidery thread.

I have also made a few roses, a small spray of blossom, and some berries all of which are also painted with a little acrylic paint then wired and bound with thread. Being fully wired, you have much more control over the manipulation of each bloom, thereby making the flowers perform for you! A flower arrangement always works better if the blooms do not all face forwards, it will appear more natural.

Materials Used:

Dupion silk in various colours, silk embroidery threads, stub wires, craft glue, artist's acrylic paint and natural tamarind fibre.

Basket of Silks

As an arts and crafts tutor and demonstrator, I am addicted to collecting colourful materials that can be used to create works of art later. I tend to get things when I see them, especially bargains, as very often when you go back for them, they would have all gone.

Here is an example of a design that I created from a selection of my silk and fabric materials collection. I chose a mixture of pink and ivory silks to blend with a few old 'Silko' cotton reels with their original cotton thread which I believe date from the 1920s.

There is quite a lot of craftwork involved in this design and it did take quite a few evenings to make all the components before it could be put together. I made silk roses using various methods, some using silk fabric that I had stiffened with gelatine.

The leaves are made from upholstery fabric that had a leaf design incorporated within it. I cut out each individual leaf and stiffened these with gelatine, then glued them onto wires wrapped with fine knitting wool.

I also crocheted a few narrow lengths of metallic embroidery silks and fine wool which I used to drape through the design to create some movement. As well as the lengths of crochet work, I made a few matching circles, some of which can be seen towards the bottom of the design attached to some fine wire mesh.

The whole design is arranged in an old picnic hamper which is resting on a willow basket. The empty hamper was then draped with a length of ivory silk fabric which acts as a background and contrast to the other materials used. I then placed a block of dry floral foam inside into which I arranged all the flowers, foliage and other components to appear as if everything is spilling out.

Materials Used:

Variety of pink shades of Dupion silk fabric remnants, upholstery fabric remnants, fine wire mesh, selection of stub wires, tube of fabric craft glue, variety of silk embroidery threads, metallic threads, touch of copper acrylic paint, knitting wool, old cotton reels, *Morus* (mulberry) bark.

Of Silk and Wool

One of my favourite pastimes is to experiment with textiles. It is the colours and different textures that blow my mind and give me lots of ideas for projects. I permanently have a few textile projects on the go which I try to fit in between everything else. The best advantage of any textile project is that it is not going to wilt or fade so you can take as long as you like on a project. If you have a busy lifestyle, it is something you can pick up and work on for a few minutes, then safely leave it for your next spare moment.

Here I have been experimenting with a couple of ideas. The first is the container or basket in which I arranged the flowers. I found an open weave wire shopping style basket from a bargain store. I experimented with weaving strips of mixed fabrics through the gaps in the wire basket which took quite a while but the finished result is great. I also wove through some scraps of knitting wool to add even more colour and fill in a few remaining gaps.

Once the basket was completed, I filled it with dry floral foam into which I could arrange the flowers and foliage.

Using a selection of silk, voile and woollen fabrics, I made a few roses attached to wire stems ready for arranging. Also, I made a few stylised flowers from glass beads carefully threaded onto fine wire.

The leaves I used have also been made from scratch using a sewing machine which was great fun to experiment with. These were made by using a free motion stitching method directly on a leaf shape drawn onto dissolvable fabric. This is a great way to experiment with an endless array of colours. Once you have finished stitching, you soak the dissolvable film in water and it completely vanishes leaving all your stitching. It is like magic! To be able to shape the leaves and give them body, I soaked them in gelatine which sets and makes them rigid and less likely to unravel. I then glued them onto wire stems to enable me to arrange them.

Materials Used:

Scraps of woollen cloth, silk and voile, knitting wool, stub wires, glass beads, metallic embroidery threads, assorted cotton and polyester sewing machine threads, and tube of 'UHU' general purpose glue.

The Weaver's Window

It can be quite a challenge sometimes when it comes to deciding the colour theme of a design. Generally, colour is one of the first elements of design that you notice; that is why it is so important to have an interesting colour combination that catches your eye.

The colour inspiration for this design was the length of woollen fabric draped towards the centre of the arrangement. I feel that this fabric has a super colour combination but it was a mission to find other materials to match and compliment. For a few days I carried a sample of the fabric around with me so that if I spotted something while flower and plant hunting, I could see if the colours would work together.

I visited a few local garden centres and plant nurseries and found a variety of plant materials in suitable colours. I was amazed at the variety of colours of summer bedding plants now available so you will always find something to suit. It is quite easy to tuck a small potted or plug plant into a design to introduce colour and interest.

This design has also worked out to be quite eco-friendly as I was able to use the absolute minimum of floral foam. I even used an old piece of recycled floral foam from a previous design. If you are not using too many stems of plant material, and you can manage to push the stems deep into the bottom of the foam, this will be fine as the water always drains towards the bottom and the end of the stem is less likely to be in an air pocket created by previous stems in the recycled foam.

The piece of spruce to the rear of the design is held in a hidden container of recycled floral foam and the central grouping of *Allium* and *Dipsacus* (teasel) are also arranged in recycled soaked floral foam. The rest of the arrangement is constructed from potted plants that will be later planted in my flower border.

All the props, dried plant material, knitting yarn and fabric were loosely placed without any fixings, glue or wire. This enables the design to be easily dismantled and reused in other projects.

Materials Used:

Heuchera 'Plum Pudding', *Nemesia* Framboise ('Fleurfram'), *Dryopteris affinis* 'Pinderi' (male fern), *Osteospermum*, *Allium*, *Dipsacus* (teasel), fresh carpet moss, spruce, larch cones, dried artichokes.

The Carpenter's Workshop

This is one of my favourite designs as I enjoy working with colour and texture so much. In floral design, colour is so important as it performs in so many ways. Colour can be used to catch your eye, portray a mood, or blend in with your surroundings as I have done here. I did not want this design to stick out like a sore thumb, just to complement the surrounding colours and tones. I wanted the viewer to look, and then look again, the more you look, the more you see!

Most of the components in this design are dried leaves and even the main flower is made from dried plant material. I kept a printed photograph of the carpenter's workshop to hand, carefully mixing colours and painting the dried leaves and in this way, I could make a design where all the colours matched and toned in with the surroundings.

Once I was happy with the colours of the plant material, I then set about wiring everything to enable the design to be arranged in dry floral foam. It is usually best to arrange flowers in situ if possible, that is the norm, but there will be odd occasion when you cannot so these I arranged at home and took them to the venue ready-made. If you are making designs this way, it is so important to take lots of accurate measurements, so then you know the finished article will fit perfectly.

To make the arrangements stand out a little, I gently gilded some of the leaves with Treasure Gold Wax (Copper colour). I thought it would be fun to include a few cobwebs and spiders in this design as you usually see them in old tool sheds. These spiders are special though as they have spun golden webs!

Materials Used:

Dried *Magnolia grandiflora* leaves, dried palm leaves, seagrass spheres, assorted dried plant material, copper pot scourers, and jute string.

In the Forge

What an amazing place to be able to arrange flowers. Who would have thought that you could possibly put a floral design in a blacksmith's forge! It just goes to show that flowers can look great almost anywhere.

The forge was a perfect setting for a design as it is a pretty much neutral colour of greys, browns and black and therefore, if you place anything of a bright colour, it will create a strong contrast and stand out.

Here I have manipulated a cheap pierced metal lantern designed to house a candle. Being made of thin metal, it was rather easy to cut slits with tinsnips, and then peel back with a pair of pliers.

Afterwards, I spray painted the whole ensemble with a grey metal primer, beforehand painting with artist's acrylic paints and then I wired onto the framework a few almost dried *Aspidistra* leaves. (Being almost dry but not completely, they still had enough flexibility to enable them to be knotted and twisted.)

These were then painted and highlighted with a little silver paint. Inside the cavity, I placed a small piece of soaked floral foam, just large enough to hold a few flowers in bright colours.

Materials Used:

Aspidistra elatior leaves, orange *Gerbera,* orange roses, cerise pink roses, orange and pink tulips, and *Ornithogalum* (chincherinchees).

The Blacksmith's Workshop

To continue the theme of the blacksmith's forge, I thought that this would be a great place for a design. Anvils are such a fascinating shape and have a long working history. Just think of all the useful objects made on this one over the decades. (I bet it has never had a flower arrangement placed on it before!)

The composition is not that complicated as there are few components used. It is made on a length of bark with a beautiful piece of driftwood glued on top to which are added a few dried leaves and some roses made from dried and painted orange peel. The driftwood was painted to simulate charcoal with a touch of gold, just enough to catch the light.

To finish the design, I simply draped silver-grey lengths of crystal effect Christmas decorations as I wanted to give the overall appearance of a recently forged work of art emerging from the glowing embers.

Materials Used:

Length of bark, driftwood from a pet shop, dried leaves from a *Protea* plant and roses constructed from dried orange peel.

Forged Flowers

It is possible to make flowers out of almost anything with a little imagination and experimentation so here I made an assortment of summer-style blooms from recycled steel and aluminium. The large rose-type flower is made from baked bean tin lids, cut to size with a pair of tinsnips and bent into shape with needle-nosed pliers.

The large flowers are cut from aluminium sheeting which is a much softer material to work with and I have made the rest of the flowers with soft drinks cans which can be curled into shape.

Once all the petals were cut out and shaped, the next stage was to assemble them, which was quite tricky as metal is difficult to stick together without welding or soldering but I managed to hold everything in place with a very strong metal glue and floral pot tape.

The centres of my flowers are made from pieces of copper pot scourers rolled into small balls and glued in place. To colour the flowers, I first sprayed them with a metal primer, then used a selection of coloured floral spray paint, and left them to dry and harden overnight, after that I over-painted with watered-down artist's acrylic paints.

Last of all, I applied a quick coat of pearlescent spray paint to help catch the highlights. Once I had finished making all the metal flowers, I arranged them in a block of dry floral foam which was held in the vice.

Materials Used:

Soft drinks cans, baked bean tins, thick aluminium foil trays, copper scouring pads, thick florist's stub wires, strong metal glue, floral pot tape, metal grey primer spray paint, coloured floral spray paints, artists' acrylic paints and floral pearlescent spray paint.

Chapter 2
In My Secret Garden

Gardener's Delight

When I am not teaching or demonstrating floral design and crafts, I do a bit of gardening for a few friends and myself. Gardening and flower arranging go hand in hand, and you cannot really make a good design without a selection of garden plant materials. There are now so many wonderful plants available, many hundreds of varieties that are perfect for cutting to use in floral designs.

Foliage is so important; it is an integral part of good design and one of my favourite foliage is the leaf of the *Heuchera* plant. There are many varieties in a huge variation of colours and sizes. They condition well and importantly last once arranged.

Here I have assembled a collection of garden tools and containers into a column design incorporating a couple of circles, one being an old riddle, and the other using willow stems loosely twisted together. I have grouped the plant material as this gives much interest and visual impact to a design.

This style of design will always need a stable stand with a strong broad base. The more weight you add, the more unstable it will become. The best way is to keep the weight central, so it will not overbalance and fall. I have used a strong metal upright welded onto a base and all the components are securely wired onto chicken wire scrunched and wrapped around the central upright hidden from view.

Materials Used:

Eremurus (foxtail lily), *Antirrhinum, Rosa* 'Peach Avalanche', *Visnaga daucoides, Hylotelephium* syn. *Sedum spectabile,* honeysuckle, rose foliage, *Gerbera, Heuchera* leaves and moss.

A Garden Riddle

There is always something pleasing and satisfying about circular designs. They always look self-contained, neat and tidy, and can also be hung almost anywhere.

This circular design is arranged in a garden riddle, which is the larger of the two, and an old flour sieve. The sieve sits quite happily inside the riddle creating a crescent space towards the left-hand side into which I could add some flowers.

I wired the sieve and riddle together temporarily with a few stub wires pushed through the mesh and twisted together at the back. I kept the wires out of view as they would otherwise detract from the design.

The mechanics of this design were quite simple, I cut some soaked floral foam to part fill the crescent void on the left, and a smaller piece of foam resting just inside the bottom of the flower sieve. To prevent the foam from falling out, I pinned it through the back of the riddle, through the mesh of the sieve, and into the wet foam with stub wires bent into horseshoe shapes. This was the easiest way as I could not get any floral pot tape through the mesh and around the wet foam.

The next step was to start arranging the plant material. I added the four *Rubus tricolour* leaves first, then some of the roses. Because there was limited space to arrange the crescent shape, I started adding the flowers at the bottom and slowly worked my way around to the top left. I then arranged the flowers in the bottom of the flower sieve to balance the design.

As this design was viewed and photographed at eye level from a standing position, I arranged it in situ. If you do not arrange your flower designs with a view of their finished resting place, you may get the angle of your plant material wrong and spoil the overall effect of the design.

Materials Used:

Rubus tricolour, *Alchemilla mollis*, lilac, *Rosa* 'Aqua' and *Rosa* 'Memory Lane'.

Cascading Cans

Galvanised watering cans are heavy as you can imagine even when empty but do make fantastic containers to arrange flowers in. What a challenge to hang eight watering cans and fill them with sunflowers.

I borrowed these beautiful old watering cans from a friend thinking they would make an unusual hanging design. It took a long time, even with the help of a friend to hang them all in place. They were difficult to suspend at the right angles and they would also constantly swing around. With the aid of string and wire, we managed to arrange them with a small watering can at the bottom and the rest with their spouts radiating outwards.

The arranging of the sunflowers was the next challenge. Sunflowers are a hefty flower and need to be held firmly in place. The best way to do this was to fill some of the cans with soaked floral foam which gave a stable base to hold the sunflowers in place. I had to be careful as the weight of each sunflower I added upset the balance of the cans so there was a lot of adjusting to be done.

After a couple of hours, the design was completed, and I was pleased with the finished result. Who would have thought such a simple idea would have been such a challenge?

Materials Used:

Helianthus (sunflower).

Vision in Blue

Terracotta flowerpots make brilliant containers for arranging flowers as they come in a wide variety of shapes and sizes. Being made of a porous material they are extremely easy to decorate especially with paint and paper.

This large terracotta flowerpot was painted firstly with three coats of off-white artists' acrylic paint. I used a large soft paintbrush to apply the paint as this helped eliminate visible brush strokes in the paint as I wanted a smooth finish. You will find that good quality paint always covers better than budget paint.

Allow the paint to dry overnight and if you notice any blemishes or brush hairs in the dry paint, you can always lightly sand them off with a very fine sandpaper and paint over.

I then decorated the painted pot with a Spode china-style design using some old, printed paper napkins. I carefully tore sections of the napkins apart and glued them on with a 50/50 mix of wallpaper paste and PVA glue. (It is always best to tear the paper when you are creating designs like this as a torn edge will not show as much as a cut edge.) So that the applied paper did not look too bulky, I separated the napkin and used single ply.

When the glue was dry, I painted the whole pot with three coats of clear matt water-based varnish to protect the decoration. I also varnished the inside of the pot to protect it from moisture as terracotta is extremely porous.

Before I added the wet floral foam, I firstly lined the pot with cellophane as flowerpots also have a hole in the bottom. Alternatively, a smaller plastic pot could be used as a liner. The wet foam was simply wedged in place with about 5 cm protruding above the rim of the pot. Once the foam was in place, I added the mixed foliage and flowers to complete the design.

Materials Used:

White spray roses, white *Phlox*, *Visnaga daucoides*, blue *Hydrangea*, garden *Clematis*, fresh lavender, *Alchemilla mollis*, mixed *Hosta* leaves and *Heuchera* leaves.

The Garden Room

This design was made on the broad window ledge of this stunning architectural garden shelter built in a contemporary style garden. The shelter was inspired by a 'Thunder house' built in a nearby Gertrude Jekyll garden. The large open space of this arch with a wonderful view of the garden beyond was an ideal spot for a few flowers.

Here I decorated part of the open space with a design incorporating a circle as the dominant feature which echoes the shape of the rounded arch above. I also used some vine prunings and tulips to give movement and interest. The colour of the tulip and matching rose was used to complement the driftwood and prunings.

To add some width to the design and make good use of the available space, I included a few lengths of *Eucalyptus* bark. The whole design is arranged in a large plastic flower trough saucer filled with floral foam. All the prunings and bark were pinned into the foam to hold everything securely in place.

Materials Used:

Tulipa (tulip), *Rosa, Begonia rex* leaves, *Wisteria* prunings, *Parthenocissus* (Virginia creeper), *Corylus avellana* 'Contorta' (contorted hazel), *Eucalyptus* bark, pears, *Magnolia grandiflora* leaves (glycerined) and *Plagiothecium undulatum* (flat moss).

Fresh for the Table

I always feel that tables look bare without a few flowers to liven things up. Flowers bring cheer and happiness whatever the occasion may be.

Here I have taken the opportunity to really go to town and show a colourful variety of seasonal flowers that create an extra special display using a harmonious colour palette.

To begin, I started by placing an upside-down wooden box with a galvanised bucket on top filled with freshly soaked floral foam in the centre of the table. This enabled me to introduce some height to the design. I then placed a few shallow dishes of floral foam around the wooden box and one underneath the table.

I began arranging with the foliage to mark the boundaries and the overall shape of the design. I find that you do not always need huge quantities of foliage within a design but you do need a variety of interesting forms to build the 'Scaffolding' of a good design. Once the foliage framework was in place, I started adding the flowers. I generally place linear forms first so here I used the larkspur.

Next, I added the largest blooms which were the beautiful hydrangea, followed by the sunflowers. Once all these blooms were in place, and being happy with the focal points, I infilled with the roses and then the finer textured flowers.

Overall, I chose a loose country style of arrangement for this traditional metal garden table which suits it rather well and portrays a relaxed, free-flowing ambience. I know that generally, you would leave plenty of space on a table for plates, cutlery and glasses but in this instance, I thought a full table of colourful flowers would give a lot more impact. This result attracts the eye and makes a conversation piece.

Materials Used:

Helianthus (sunflower), *Hydrangea, Phlox, Rosa* (rose), *Visnaga daucoides, Alchemilla mollis* (lady's mantle), *Lavandula angustifolia* (lavender), *Consolida ajacis* (larkspur), *Bupleurum fruticosum* (shrubby hare's ear), *Ballota pseudodictamnus* (false dittany), *Clematis* trails, *Hedera* (ivy) and *Bergenia cordifolia* (heart-leaf bergenia).

The Water Sculpture

This elegant design was made on the grounds of a magnificent garden that belongs to good friends of mine who very kindly allowed me to borrow their pond in which to site this arrangement. It was not easy as the design itself is standing in 46 cm of water, so the only way to arrange the flowers was to stand in cold water wearing a pair of waders.

The main structure was assembled from pieces of driftwood wired and screwed to a metal stand. This was made beforehand, and then placed in the pond. The white tulips were all inserted into small plastic orchid tubes containing a little water which had been hidden. On top of the design, I placed some soaked floral foam in which all the other flowers were arranged. The columns of cascading hyacinths were made of single floret pips threaded onto fine wires (i.e., they were pipped).

Also hidden amongst the flowers are small tubes in which water is pumped from a submersible pond pump in the actual pond itself. I was very careful to hide all the tubing. Once the pond pump was switched on, the streams of water completed the composition.

This would make a great design for a garden party or a wedding.

Materials Used:

White tulips, white hyacinths (Pipped), white *Ranunculus*, white *Syringa* (lilac), *Visnaga daucoides* and a little fresh moss.

The Potting Shed

There is nothing like a tidy, organised place of work, especially a potting shed. This potting shed is in a friend's garden and is always well kept. It has a tool for every job, and plenty of space for potting and everyday jobs needed for a busy working garden.

This design is a large still life and so I needed to rearrange some of the tools and boxes and introduce a few props to add to the ambience. I hung some dried flowers and placed some stems of birch and hazel here and there, including a large bunch of dried *Hydrangea*.

As this was photographed in late spring, I used a selection of seasonal bulbs and bedding plants to add a splash of colour and to emphasise the focal area which is the potting bench. I placed a hand-tied bouquet of blue hyacinths and white *Ranunculus* tied with natural raffia.

Materials Used:

Salix (pussy willow), potted *Narcissus,* potted primroses, blue hyacinths, white *Ranunculus*, daffodils, tulips, potted *Ranunculus*, *Wisteria* prunings (wreath on wall), Corylus *avellana* 'Contorta' (contorted hazel), dried sea lavender, dried *Hydrangea,* apples, and dried roses.

The Flower Clock

Whenever I find some spare time, I manage to do a little gardening for a few friends. One garden I look after had a small circle of grass that never really grew very well and was always boggy in the winter. So, one summer the owner and I decided to dig up the lawn and replace it with gravel and an ornamental stone circle in the centre. After the stone circle had been installed, I thought that it would be a good spot for another garden design.

So, I set about making this arrangement surrounded with tea lights. The centre of the design is arranged in a 41 cm posy pad filled with an array of pink and white flowers with a touch of lime green. The lime collar brings out the pink tones in this design.

The large badminton racket-shaped petals arranged around the outside are made from a roll of woven natural plant material mesh. This was cut out into petal shapes, and then had rustic wire glued around the edge for support. I then stuck a few spirals and swirls of rustic string onto the panels, also a few pieces of sisal at random.

When the glue was dry, I painted the panels with a mixture of spray paint and artist acrylic paints in pink and pearlescent colours. I left the painted panels a couple of days to dry completely, and then attached some stout stub wires to make a false stem so that they could be inserted into the posy pad.

To complete this design, I added some tea lights around the outside, being careful so as not to catch the design on fire. I think that this design would make a great centrepiece for a garden party!

Materials Used:

White *Agapanthus*, pink *Phlox*, *Rosa* 'Blueberry', pink *Veronica*, white *Trachelium, Alchemilla mollis,* lime green sisal.

A Floral Candelabrum

Designs always benefit from movement, leading your eyes around, and looking at all points of interest on the way. Starting at the focal point or dominant area which here, in this design, are the lime green *Anthurium* blooms, your eye slowly travels upwards and follows the curve of green orchids around the rustic willow wreath ring. The *Anthurium* flowers draw your gaze back to the centre to dwell on the flowers there where the shiny texture complements the rough twigs and branches.

This arrangement is made of two main components; one is a simple framework of ash tree stems. These were wired securely together and then pushed into the ground. Towards the top of the framework, I wired a pussy willow wreath ring onto which the orchids were attached.

The second component is the main arrangement at the base which is arranged in a hidden bowl of floral foam. A very useful foliage is the *Aspidistra* leaf. It is very robust and can withstand being manipulated (Woven, stapled, twisted, rolled, and even painted). Here I have rolled some, while a few have been twisted, rolled, and wired in place. Very often they come dusty and usually need a wipe with a damp cloth.

Materials Used:

Lime green *Anthurium andraeanum* 'Midori', *Aspidistra* leaves, lime green carnations, lime green Singapore orchids, bear grass, *Salix* (willow) and *Fraxinus* (ash) stems.

High Tea

Flowers always add a little magic to any occasion, especially a friend's birthday. I always jump at the chance to arrange a few flowers whatever the occasion and this was a good opportunity to experiment with colour and have fun. For this event, it was Willow's birthday, so we invited her friend Tolly for tea.

I found a good spot on the lawn and set up a picnic. I placed a few stout boxes under a tablecloth to raise the design off the ground. This helps to prevent the overall design from appearing flat and all on the same level.

Next, I placed the main props i.e., the hamper, jug, cakes, cup and saucer and teapot with plenty of space between them to allow room for the flowers.

Once I decided where the flower placements were going, I added trays of freshly soaked floral foam and arranged the flowers in situ. I used the sunflowers as the focal flowers which were distributed throughout the whole design. I then arranged the other flowers in a mixed fashion throughout including a few in the jug placed on top to add height. I pinned a rose garland around the top of the tablecloth which follows the perimeter of a large wooden board underneath used as a tabletop.

Lastly, I invited our special guests for their special birthday treat. On hot days, it can be a challenge to make flowers last, so it is always a good idea to check the weather forecast and choose flowers to last accordingly. You will need to water the containers and trays well and mist regularly with a water mister. (The dogs enjoyed the cool water mist too.)

Materials Used:

Chrysanthemum 'Shamrock' bloom, carnation blooms, *Alchemilla mollis*, mixed spray roses, large *Helianthus* (Sunflower), mini sunflowers, *Danae racemosa* (Soft ruscus), *Dahlia* 'Cornel', *D.* 'Garden Festival', *D.* 'Bargaly Blush', *D.* 'April Heather', *D.* 'Westerton Folly', *D.* 'Christopher Taylor', *D.* 'Hillcrest Candy'.

Mid-Summer Pedestal

It is always a great pleasure to be able to arrange a pedestal design especially a large one with fantastic plant material. A pedestal design is basically a triangular-shaped arrangement which is generally one of the first shapes in floral design most people will be taught to make. It is such a useful shape to get to know as it has many uses at various functions.

It may appear a simple shape to put together but there is more to it than meets the eye. It is very tempting to add too many flowers and make them appear solid and stuffed. You must leave some space and place a few flowers deeper in to give the appearance of depth.

This design is arranged in a jumbo-sized block of floral foam sitting in a large plastic plant bowl which will hold plenty of water. You will always need to make sure that your wet foam is securely held in place with a good quality pot tape (double thickness if you can) as pedestal designs are generally heavy and contain lots of plant material.

Materials Used:

Gladiolus, Delphinium, Matthiola (stock), *Paeonia* (peony), *Hydrangea, Astilbe, Asclepias, Bouvardia, Amaranthus,* David Austin *Rosa* 'James Galway', *Rosa* 'James L. Austin', carnations, *Elaeagnus* 'Quicksilver', *Hosta* leaves, *Rubus tricolor,* white oriental *Lilium* (lily)*, Ballotapseudodictamnus, Polystichum* ferns, *Physocarpus opulifolius* 'Diabolo'.

Chapter 3
Coastal Inspirations

The Seahorse

A good piece of driftwood is such a useful object to have if you do lots of flower arranging. It is not only beautiful but it's amazing natural form also offers so much interest to any design, even before you start to add flowers, and where do you add the flowers exactly? How can you enhance a piece of nature? I think that the best way and starting point is to study your piece of driftwood. Try standing it in different positions, different angles, and different light, and do this over a period. Do not rush into a design right away, especially when using driftwood. Look to see if the way it is standing is visually balanced, or could you balance it with some plant material? Do not worry if it will not physically stand how you want it as you can quite simply screw a wooden base in place so that this can be achieved.

When you have decided which way up it should be and made it stable, the next job is to decide where to add the flowers, thinking about the balance of your design. It is very easy to make a design look unbalanced or top heavy. You could think of your design as a set of scales or a seesaw. By adding material to the right location, it should sit comfortably. In the wrong location, it will visually tip the scales and unbalance everything.

You have decided on the right place for the flowers; the next task is to sort out your mechanics. This can also be tricky as you will need to find a way of securing some floral foam to the driftwood. Try to use the absolute minimum amount, as the more you use, the harder it will be to disguise. Remember the mechanics of a great design should be cleverly disguised to be virtually invisible.

I kept the flower design very simple and limited by balancing the driftwood with a flowing line of pure white *Anthurium* blooms. To achieve this, I used mixed sizes of *Anthurium* which I graduated from large at the top to the smallest at the bottom. To create the sweeping line, I wired each *Anthurium* in turn like a garland and attached this to the top of the driftwood by wiring it to a cup hook hidden from view.

To add some visual weight to the bottom of the design, I added some rope, seaweed and a few beach pebbles, which also hides the plywood base. To help continue the downward movement of the *Anthurium* line, I incorporated a trail of crocheted shredded *Phormium tenax* (New Zealand flax) leaves.

Materials Used:

Driftwood, pieces of dried seaweed, rope and twine, dried *Xerophyllum tenax* (bear grass), crocheted *Phormium tenax*, *Anthurium* blooms.

Sails

What a challenge this design was! It was very enjoyable to set up but quite a task as we were working against the tide and the weather.

The two sculptures were made in advance in the comfort of a warm kitchen and then transported to the beach. I constructed the frames from galvanised fencing wire about the thickness of a wire coat-hanger. I fastened the wireframe together with strong floral pot tape then bound the main skeleton with garden string. Once the main frame was completed, I then wired on the swirls and spirals of 'Rustic Wire' (A floral sundries product from OASIS® of flexible wire wrapped with dried plant fibres). I also used paper-covered tying wire of which the ends were curled around a BBQ stick.

I then glued a piece of hessian fabric inside the spathes to form a stable surface onto which I could glue flowers.

The flowers are the individual cut flower heads of beautiful blue *Delphinium* and tiny wired bunches of *Gypsophila*. All of these were then glued onto the hessian backing with 'OASIS®' Floral Adhesive. Because the flowers were not in a water retaining medium, they had to be sprayed regularly with water and kept in a cool dark place until ready for photographing. (This style of design can only be made a day in advance.)

These types of designs are also very time-consuming and you might need help especially when there is a lot of wiring involved.

The completed sculptures were stood on heavyweight metal bases and partly buried in the sand. The final job was to wait for the tide to come in and submerge the stands.

Materials Used:

Delphinium and *Gypsophila paniculata* 'Million Stars'.

The Beachcomber's Bounty

There are so many interesting places along the coast where a design could be made. It is probably just me; I am always spotting niches, tops of walls, platforms and structures where a few flowers could be placed for a photograph.

This wild-looking coastal design was not quite arranged where I first planned to make it. On a previous visit to check out the area to plan a design, the weather was perfectly calm and sunny. When we returned to arrange and photograph this design, the weather was very windy. So, plan 'B' was put into action and this sheltered spot was chosen. (You can just make out the rough sea over the top of the wooden groyne)

The mechanics of this design were very simple to put together, just a long tray of soaked floral foam which I temporarily taped in place on the wooden beam. It needed to be securely fixed as I used lots of heavy plant material within the design which could cause it to easily tip forwards.

The next step here was to arrange all the dried seaweed as this is the main feature of the design. To make the seaweed more interesting and stand out, I painted it this unusual apricot colour and splattered it with acrylic paint using an old toothbrush. Once the seaweed was in place, I could arrange the rest of the materials. To give a wild effect with the calla lilies, I kept the stems long which are naturally lime green and lifts the design.

The bell cups, which are natural seed pods, were also painted to blend with the colour theme and add interest with their volumetric form. All the other flowers including the orchids were then placed radiating from the central point.

To complete the design, I added a little sisal fibre which I also coloured with peach fabric dye so it would blend with the rest of the materials. This gave a soft contrasting texture and filled any unwanted gaps.

Materials Used:

Dried seaweed, bell cups, sisal, *Vanda* orchid, *Zantedeschia* (calla lily), *Leucospermum* (pincushion protea), *Morus* (mulberry) bark, *Mokara* orchid.

Caught in the Breakwater

A colourful object will always stand out from its surroundings if the background is of a neutral colour. Here I wanted to create something so bright it would almost jump off the page.

This design is a swag arrangement constructed from a combination of colourful materials I have collected and some flowers that I have made especially for this design.

The main structure of the swag is a length of bright lime green natural fibre mesh entwined with chunky lime green woollen cord and a piece of decorative netting. I chose to use lime green for the background as it will complement any bright colour you put with it and make that secondary colour appear even brighter and really stand out.

The large orange flowers I have made by 'Wet-felting' merino wool into a square, then cutting petal-shaped slits into the felt and stiffening with gelatine. It is quite a long process but worth the finished result. Using the gelatine makes the felted flower more robust and helps it to keep its shape. The centre of the flowers is created from ready-made cake decoration stamens which are readily available in a vast array of colours.

The two pink string style flowers are made from garden plant ties which are natural jute cords with a flexible wire through the centre. These plant ties can easily be cut to any length and shape. These have been painted with artists' acrylic paint which will also help to stiffen the flowers and hold the petals in place. The centres of these pink flowers are made from sisal fibres rolled into a ball and painted with acrylic paint.

The lime green daisy style flowers are made from florists' paper-covered wire that has also been painted. The pink tendrils are made from stub wires wrapped in mulberry bark fibres which have been dyed pink. The fibres are held in place with UHU glue.

Last of all, there are a couple of deep pink rose-style flowers made from fine mesh style ribbon which I have also painted. You do not always need to worry or be exact when buying materials to construct flowers as virtually all or most components can be coloured, painted or spray painted in some way. The whole design was assembled with wires to hold everything in place securely without the need for too much glue.

Materials Used:

Merino wool, ready-made flower stamens, *Morus* (mulberry) bark, paper-covered wire, corded plant tying wire, woollen cord, assorted decorative mesh, stub wires, mesh style ribbon, artists' acrylic paint.

Under the Sea

Life under the sea is bursting full of colour and beauty. There are so many amazing formations, especially coral. So, here is my interpretation of a coral reef, deep under the sea, somewhere in warm tropical waters.

The first thing I made here was the coral reef itself. I started with a sheet of loft insulation eco board. This is available from most builder merchants and is a great material to use as it is very easy to cut, shape and glue together.

So, once the main shape of the reef was complete, I coated it with a layer of silver sand mixed with PVA glue and acrylic paint, which should be applied with a small palette knife. This will take 24 hours to dry.

The next job was to add all the strange but beautiful things that grow under the sea. I used materials such as dried bracket fungus upside down as it looks very much like coral. This was coloured with acrylic paint. I also used a piece of driftwood that was hand-painted to transform its colour.

The fish were great fun to make but a bit fiddly. The bodies were carved from the same eco board as the coral reef, and then covered with paper fish scales all glued on individually. (Be careful what glue you use on the eco board as solvent-based glues will dissolve it. I found that latex-based glues and PVA glues are fine.) These were also finished with acrylic paints and a spray of pearlescent glitter spray paint.

Materials Used:

Dried bracket fungus, driftwood, air plants, small lotus pods, dried *Scabiosa stellata* 'Paper Moon' seed heads, *Celosia argentea* var. *cristata*, *Chrysanthemum*, skeletonised leaves.

Flotsam and Jetsam

There is nothing like a stroll along the beach to blow away the cobwebs, especially if you are feeling a little low. The sea air is good for you. I like to visit just after there has been a storm or high tide, as amazing things are often washed up. Some objects may have travelled for hundreds of nautical miles. Very often there will be an exciting tangle of coloured seaweed entwined with driftwood or fishing net.

Here is my interpretation of an entanglement washed up after a high tide. The main component that I have used is not seaweed but silk and cotton embroidery threads that have been tangled and hardened with fabric stiffener. It really does take on the appearance of brightly coloured seaweed. The choice of colours is endless, and you can come up with no end of combinations.

Once I had made the entanglement of stiffened threads and they had set hard, I set about collecting an assortment of plant material and seashore finds. Again, you will need to experiment with this type of design over a period by trying different materials, colours, and textures together. A focal point is still needed which in this case it is the sea urchin.

Also, some gentle movement needs to be included to prevent all the fussy textures from becoming overwhelming. This design would look great as a table centre entwined with a few clear fairy lights.

Materials Used:

Assorted silk and cotton embroidery threads, fresh bear grass, spent rose heads, *Limonium, Sedum rupestre* 'Angelina', variegated *Sedum*, seashells, small pieces of driftwood, glass beads and sea urchin.

The Spider Crab

Here is another example of a design incorporating driftwood. I've quite a large collection of pieces of driftwood now that I have accumulated over the years. Many pieces have been given to me by fellow flower arrangers who have rung to say, "I'm clearing out my shed and I've got a nice piece of driftwood that you might like."

This arrangement is made on an old tree stump. To give you an idea of scale, it is around 1.2m at the widest part and very heavy too. I have used the tree stump this way up as if standing on legs to give a feeling of a creature from the deep.

On the top of the stump, I glued on a piece of dry floral foam using a cool melt glue gun. In the foam, I have arranged two large exotic-looking flowers that I have made with some sisal sheeting and coloured knitting wool which have been painted to complement the *Hydrangea* flowers and bear grass. The tangled form towards the right of the design is also dried bear grass that I have knotted and woven together using a crochet hook and painted to match all the other components in the design.

The selection of seashells has been painted as well to complement but in a darker shade, to give some variation in tones of colour. These have very simply been held in place with a floral fix. I have also hung a small piece of black net, decorative vine and dried lichen from the centre.

Materials Used:

Dried *Xerophyllumtenax* (bear grass), dried *Hydrangea*, preserved lichen, flowers made from abaca sisal sheeting, and assorted seashells.

The Mermaid's Pearls

Sometimes a design takes a while to develop and evolve after quite a lot of experimentation, trial and a lot of error. I had a rough idea of what I wanted to make as I could see the finished design in my mind but the difficulty is taking that image from your imagination and turning it into a finished article that you can touch, pick up and hold.

For this design, I found an empty box and over a period of a few weeks, I slowly collected some materials including shells, wire, pearls, paint etc. When I felt I had quite a collection of items with a connection to my chosen theme, I tipped everything out onto a large table and started to experiment. I laid out the materials loosely in a circular form and tried the various items in different combinations.

So here I started with a few pieces of dried twisted seaweed, then added a couple of pieces of driftwood and some pearl beads which I had wired to give movement to the design. I then thought about a focal point or dominant area for which these small mussel shells were perfect. I found these mussel shells in a garden centre and they were wired onto a wreath frame which I dismantled to make into flowers.

To give even more movement and interest to the design, I added a few stub wires which have been wound in fine knitting wool, and then bent into shape ending with small spirals.

When I was happy with the roughly laid out design, I took a photo so that I could remember where everything was placed. I then painted some of the materials with acrylic white and pearl paints. Once the painted items were dry, I glued everything together.

To complete the design, I glued on a mixture of tiny glass seed beads, small forget-me-not style flowers made from tracing paper and tiny pearl beads. This finished design would make a perfect centrepiece for a dining table or could also be hung on a wall.

Materials Used:

Dried seaweed, driftwood, *Morus* (mulberry) bark, pearl beads, glass seed beads, string, silver wire, tracing paper, cotton muslin fabric, fine knitting wool, silver embroidery thread, and UHU® glue.

Evening Glow

On a previous visit to my favourite beach, I could not help but notice the beautifully weathered wooden steps that go down to the shoreline. The salty sea air and rain have rendered them as smooth as silk and turned them silver grey in colour. I thought that this would be a fantastic place to stage a design that would sit comfortably while cascading from top to bottom with lots of gentle flowing movement.

I decided that it would be a good idea to make this design and photograph it in the evening when there was hardly anybody around so nobody could trip over the flowers on their way to the beach! Also, the amazing evening light gave a gentle warm glow to the whole scene.

I chose to use a rough texture that would contrast greatly with the smooth wooden steps and liked the idea of rustic half spheres filled with the dried plant material that would be linked together and create the movement needed. I made these spheres by soaking garden hessian leaf sacks with fabric stiffener and then wrapping them around a balloon. Once the stiffened sacking had set which takes a few days, I popped the balloon and cut the sphere in half. They make great basket-style containers and can be spray painted any colour. Here I used a mixture of gold, copper and mauve paints. Because this was only a temporary design and consisted of dried plant material, some of the components were placed in and around the spheres loosely without the need for floral foam.

To link all the rustic spheres together, I used a couple of strips of matching mauve-coloured silk fabric intertwined with a length of crocheted dried plant material. A useful plant fibre that you can use is sisal. This amazing material is an incredibly strong fibre and is also very flexible. I have also draped a few narrow lengths of colourful woollen fabric which picks out some of the tones in the *Hydrangea* blooms and roses.

Materials Used:

Dried roses, dried *Leucadendron* 'Safari Sunset', dried *Hydrangea* (all of which have been hand painted or spray painted), sisal, scallop shells and assorted small shells, and fabrics.

The Beach Hut

I always try to find time for a stroll along the beach whenever I can to get away from the hustle and bustle of a busy life. A walk along the beach magically de-stresses you and blows away the cobwebs. It's also great fun to see what has been washed ashore as it can be absolutely anything and may trigger off some brilliant ideas.

I do have quite a collection of things I have picked up and been given over the years, and a great way to display some of these treasures is to use them in a collage.

You cannot rush a collage as it will take time to construct. The best way is to simply lay a few of your chosen pieces out on a board or tray and see if you can make an interesting design or pattern out of them.

Here I have constructed my collage on a backboard made from a sheet of plywood covered in a mixture of horticultural sand and PVA glue. Mix the sand and glue together into a sloppy mixture and then spread it all over the board. This mixture will take a few days to dry and becomes very hard and solid giving a great background on which your collage can be made. Once dry, I glued around the edge of the board some flat pieces of driftwood to form a picture frame, some of the heavier pieces had to be screwed on.

Next, I experimented with the layout of the design over a period of a few days until I was happy, and then started to assemble the items. I stuck to the general rule of floral design of larger objects towards the centre and smaller finer objects towards the outside. You still need to add movement and rhythm, so this was achieved using string, hessian fabric and seaweed.

Some of the heavier pieces had to be wired onto the board by drilling tiny holes through the object and board so a thin wire could be passed through and secured at the back, out of view. The rest of the lighter-weight pieces were held in place with a hot-melt glue gun or UHU glue.

To complete my collage, I aged and distressed some of the newer-looking objects with watered-down washes of acrylic paint and splatters. If you are going to hang a heavy collage like this, please make sure your hanging method is sound and secure.

Materials Used:

Old gardening glove, child's windmill, half a plastic milk carton, plastic starfish, nylon string, hessian fabric, dried seaweed, driftwood, dried *Eryngium* (sea holly) flowers, dried lichen, and few seashells.

The Lobster Pot

This idea began with a lobster pot which I found one day in a garden centre. This intriguing object has a fascinating form with lots of interesting shapes used within its construction. The main shape is a cylinder with lines cutting through creating a cage. It is also completely made from natural plant material which complements any flowers and foliage you display with it.

I decided to use my lobster pot laid on its side, partly buried in the beach with flowers spilling out of the centre. By burying part of the lobster pot in the shells, it created a stable and level base on which I placed a dish of soaked floral foam. This also helped prevent the lobster pot from blowing away in the wind.

This design needed a bright, dramatic colour that would contrast with the beach and stand out. So here I chose these wonderful vibrant shades of cerise pink and red. The main flower I used was the fantastic *Gloriosa* which looks so exotic and great when used in a group but does not have a particularly long stem so some of them I have arranged in water-filled plastic orchid tubes hidden amongst the shells on the beach.

I could not resist these amazing *Hydrangea* flowers as the colour was perfect to place towards the back of the arrangement to add a feeling of depth. I also found these dark cerise roses that tone with the overall colour scheme.

To give a little movement and prevent the arrangement from appearing static, I added a couple of *Phormium* leaves weaving their way through the design. I also introduced some OASIS® Rustic Wire and lengths of mulberry bark which I have dyed the same colour red to match the flowers.

To complete the design, I tucked a little sisal fibre which also went into the same colour dye to enable everything to match and blend perfectly.

Materials Used:

Gloriosa, roses, *Hydrangea*, sisal, Rustic Wire, *Morus* (mulberry) bark, *Phormium cookianum* subsp. *Hookeri* 'Tricolor'.

A Seaside Treat

Spheres are such a versatile form when it comes to floral design. This spherical form is an amazingly useful surface on which you can display and emphasise textures.

Texture is an important design principle within the floral design and more importantly so when using a minimum selection of plant materials. Texture can be used to enhance a certain area of a design or make it less dominant.

This design is made from five spheres all different from each other being various sizes and all with different textures. Four of the solid spheres featured here are polystyrene which I firstly covered with a layer of PVA glue and tissue paper. This acts as a barrier which prevents solvent-based glue or glue from glue guns from melting and eating into the polystyrene.

The top sphere is covered with dried plant fibre, and the second is covered with a mixture of builders' sand and PVA glue. Once dried, this sandy mix gives the appearance of a beach surface and creates a rough texture.

The middle sphere is covered with a sisal string which I glued on in a series of spirals overlapping each other. To take away the plain colour of the string, I painted the finished sphere with washes of watered-down acrylic paint in shades to match the birch bark used on the fourth sphere.

The fourth sphere is completely covered with strips of *Betula* (silver birch) bark carefully peeled into thin layers and glued in place with UHU® glue. This creates a wonderful rough texture with much interest, especially when you look closer.

The fifth and final sphere is a cane sphere which I bought ready-made from a garden centre. This sphere's surface is broken with holes and interesting shapes created from the construction of the woven cane. The holes make the sphere appear lighter and less dominant.

Once all the spheres were decorated with interesting textures, I drilled a hole through the centre of each one so I could slide them onto the metal stand bound with string.

If I had left this design with just a selection of round forms, it would look static without any movement or rhythm and the overall arrangement would not work. The pieces of dried seaweed I added, gives all the movement and rhythm it needs linking all the spheres together creating a harmonious design.

Materials Used:

Dried seaweed, dried *Hydrangea*, dried *Eryngium*, cane sphere, *Betula* (silver birch) bark, sisal string, jute string, builders' sand, PVA glue, glue gun and UHU® glue.

Chapter 4
Woodland Wonderland

The Magenta Tower

An unusual floral design calls for an unusual container, structure or stand. This amazing design is arranged on a raised tower construction made from an assortment of materials. The structure had to be sturdy in order to take the weight of the plant material which is mainly arranged on the top of this design.

A strong structure must start with a solid base, so here I used a double thickness of plywood glued together into which I could drill deep holes and glue in the uprights. I used stout bamboo canes covered in a mixture of sacking, moss and string which all give a rustic texture. The cross members of the framework are made from willow stems and contorted hazel, some of which are bound with moss and coloured sisal. These are the most important part of the tower as they give strength to the structure and prevent it from wobbling. All these pieces are wired together and then bound with string.

The main part of the flower design is arranged in a tray of freshly soaked floral foam securely taped to the top of the tower.

The focal point of this design which is towards the top is emphasised by a few polystyrene spheres covered with *Heuchera* leaves and then bound in purple twine that I dyed to match the colour of the plant material. I arranged the design in a cascade sweeping down from the left-hand side down towards the small placement at the base. To create the movement of plant material, I made a small length of garland on a wire which could be bent into shape.

This design was made on a particularly hot day, so all the fresh plant material had to be regularly sprayed with water to help prevent anything from wilting.

Materials Used:

Rosa 'Blueberry', *Veronica, Zantedeschia* (calla lily), *Astilbe, Nigella* seed heads, *Heuchera* leaves, and sisal.

The Stag

Woods are fantastic places full of inspiration. Every direction you look, there will be something amazing to see. If you stand still and listen, you may come across wild animals, maybe even a stag! This one I carved from large sheets of insulation foam which were glued together. The legs were strengthened with a stiff wire that is glued into the body.

Once I had carved the stag (which, by the way, made a terrible mess indoors), the next job was to cover him with moss. I used dried flat moss which you can find at the florist's wholesalers. This did take a few evenings to achieve, by glueing and pinning it in place. His nose is made from dried palm bark touched up with a little paint.

The antlers were made from strange-looking dried fasciated plant material *Crypotomeria*. This was given to me by a friend and I knew it would come in handy one day. To add a little colour and interest to the design, a few fresh flowers were placed here and there on the forest floor.

Materials Used:

Delphinium, *Astilbe*, *Hydrangea*, *Paeonia* (peony), *Eustoma* (lisianthus), *Cryptomeria japonica* 'Cristata'.

The Old Gate Post

Very often inspiration suddenly appears when you least expect it, just like a eureka moment. It usually happens when you are not really looking or thinking about a design. My inspiration for this design struck me while out on a woodland stroll when I come across an old broken fence post entwined with rusty barbed wire. The way the wire spiralled around the post was so interesting with beautiful rhythm and movement. It would be perfect if I could recreate and incorporate it into a floral design.

The fence post I used for my version was rescued from a bonfire which was about to be lit. It was ideal, an old 10cm x 10cm fence post beautifully weathered and rotten at one end. I cleaned it up with a stiff brush and then painted it with watered-down PVA glue. This helped to hold the loose surface of the wood together as some was falling away and I did not want to lose its natural weathered patina.

Next, I fastened the bottom of the post to a stable plywood base to make it stand and a second post at a 45-degree angle. I then wrapped a piece of rusty barbed wire around and held it in place with nails in the back of the post. Within the space created by the wire and sloping post, I created a cobweb with metallic embroidery thread. This was quite difficult to make, knotting the thread into radiating strands in a spiral fashion. A small dab of super glue on each knot held everything in place. One of my homemade spiders took up residence in the centre.

To increase the movement in the design, I included some sinamay ribbon which I painted green. Also, I entwined the barbed wire with some stub wires wrapped with silk waste knitting wool.

The dog rose style flowers are made from *Morus* (mulberry) bark painted with acrylic paint. I made the leaves from mixed fabric stiffened with gelatine then hand stitched with veins. These were then wired into small branches and wrapped with green embroidery threads.

The blackberries are made from black glass seed beads glued to a larger central wooden bead, also painted black.

Once I had made all these components, I wired them in place to a couple of nails in the post hidden from view.

Materials Used:

Silk waste knitting wool, embroidery threads, *Morus* (mulberry) bark, sinamay ribbon, glass seed beads, artist's acrylic paints, UHU® glue and super glue.

The Enchanted Tree

I wanted to create something magical and enchanting. Where do you start with a subject matter such as this that has a mystical atmosphere?

With all designs, there is always a trigger point that starts the ideas rolling in. The starting point for this design was a fantastic piece of old *Wisteria* stem which I rescued from a garden that was about to be sawn up and burnt. It is always a great shame when a favourite plant suddenly dies in your garden but it is not so bad when you can make use of the remains. I decided that it would make the perfect tree trunk for a magical design.

I trimmed and tidied the *Wisteria* stem and fastened it to an old metal floral stand screwed to a sturdy wooden base painted black. I felt that the main stem needed more movement, so I wired in place a few stems of old *Clematis* vine and a few smaller branches of *Wisteria*. When I was happy with the form of the tree, I painted it with spray paint, acrylic paint and glitter glue.

The leaves I constructed from painted *Morus* (mulberry) bark fibre glued and wrapped around leaf-shaped wire frames. These were then attached to the tree using long lengths of rustic wire which were also painted.

The flowers were made in a similar method as the mulberry bark leaves but consist of six petals wired together and then bent into shape like an English woodland bluebell. These were then also attached to the tree with lengths of painted Rustic Wire.

Towards the bottom of the tree, I added a large, stylised fern which I made from one-millimetre gauge purple-coloured aluminium florist's wire. This also adds to the ambience of the mystical woodland theme and introduces a little purple colour towards the base of the design.

The toadstools at the foot of the tree are made from mulberry bark but constructed using a papier mâché style method over a toadstool-shaped mould. These were also then painted with the artist's acrylic paints and a dash of glitter glue, Rustic Wire.

I really enjoyed this photoshoot as it involved using smoke pellets to create a misty atmosphere.

Materials Used:

Wisteria stems, *Eucalyptus* bark, *Morus* (mulberry) bark, *Clematis* stems, assorted acrylic paints, glitter glue and fresh carpet moss.

The Fairy Ring

It is great to be able to make something that is completely different and out of the norm. I really went mad with this design in the woods. We carried buckets of flowers and soaked floral foam into the woods where we found this quiet spot, free from dog walkers and riders on horses. (Shame about the midges!) I arranged about eight or nine plastic trays holding soaked floral foam in a circle, making sure that the woodland floor was level to enable the trays of foam to sit flat and stable.

Next, I arranged a few fallen moss-covered branches around the perimeter of the circle. I then started adding flowers, mostly at random to give the impression they were magically growing in a circle. The shepherds' crooks were made from coat hanger wire bound with pink ribbon and string.

The round lanterns were found in a local garden centre and simply sprayed to match. After adding most of the flowers to the floral foam, I placed some into plastic orchid tubes containing water, and then inserted these directly into the ground.

Please be careful when painting objects that will likely encounter a naked flame, such as candles. Not all paint is fire retardant! I would suggest that you use battery-powered LED candles. We took every precaution while using tea lights in the forest.

Materials Used:

Pink roses, pink *Astilbe,* lilac *Delphinium,Polianthes* (tuberose), cultivated *Digitalis* (foxglove), white *Nigella* (love-in-a-mist), mauve *Phlox, Alchemilla mollis*, and a few *Alyssum* (summer bedding plants).

The Woodland Bridge

One of the hardest things I found producing this book was choosing a location that lends itself to becoming a setting for a floral design. There are so many places and some very difficult to get to, especially tranquil woodland settings such as this. You always end up parking so far away from the chosen spot. But no matter, it is a delight to come upon that special spot.

This beautiful shallow woodland stream has been bridged with an arched design. It has been made on a sturdy wire frame which has been pushed into the bank, on either side of the stream. Luckily, the stream was only a couple of inches deep so there were no worries of water in the wellies.

The wire frame was covered and bound with fresh moss, some artificial lichen, and a few stems of basket willow. Tucked here and there, I hid a few pieces of soaked floral foam and a few plastic orchid tubes filled with fresh water. These components have been carefully camouflaged with more, fresh green moss.

I then carefully arranged the flowers in a random fashion but keeping it balanced so as not to overpower any spot specifically. I chose to use the colour orange as it contrasts well with the natural green surroundings, enabling the design to stand out and be seen.

Materials Used:

Orange *Zantedeschia* (calla lilies), orange roses, *Asclepias tuberosa,* yellow spray roses, *Dendrobium* orchids, *Amaranthus,* string of hearts, green primrose, dried *Muehlenbeckia* stems and fresh moss.

Fallen

This is such a beautiful piece of driftwood I had to use it and thought that it would be ideal for a woodland design. This was a piece that I bought at a sale from a flower demonstrator who was having a declutter, so every time I use it, I am reminded of an old friend who makes beautiful designs also.

This driftwood is so full of interesting undulating forms and fantastic textures; it is really the star of the show. I felt that this design did not really need a huge variety of plant material, just a few items and an interesting background and foreground to complement.

Full of cracks, crevices and holes it had lots of places to insert small water-filled orchid tubes into which I could add a few fresh flowers.

The main flower featured here is *Visnag adaucoides* which I chose for its semi-flat form with a slightly rough but delicate texture. It also sits comfortably within the driftwood and brings great interest to the design. My second choice of flower is a few green chrysanthemums which give a fresh feel to the arrangement and add some colour.

To include a contrasting texture, I added a few lime green *Hypericum* berries which I have removed from their stems. These are long lasting even when removed from their source of water.

To add further interest, I placed the whole design in a complementary surrounding which creates a Rustic woodland ambience. The background and foreground are decorated with fresh carpet moss, variegated ivy and a few dried, curled-up leaves which complement the design and repeat the movement and rhythm of the driftwood.

Materials Used:

Hypericum berries, spray chrysanthemums, V*isnaga daucoides* (bishop's weed), carpet moss, variegated *Hedera* (ivy), dried *Quercus* (oak), *Fagus* (beech) and *Acer* (maple) leaves.

Butterflies

I was not sure where to start with this design as it was so different. I do like creating unusual fantasy-style flowers and plants that could easily have come from another solar system! It is great just to let the mind go and do some daydreaming.

I wanted this design to look like a strange tropical plant bursting into bloom. The large green leaves at the base are made from thick wire mesh cut into leaf shapes and covered with dried palm tree bark, jute twine and sisal. Most of this was glued on with my trusty low-melt electric glue gun.

Next, I simply painted the leaves with acrylic paints before they were then pushed into the soil forming a circle around a plastic bowl containing floral foam. Into the top of the foam, I added a few stems of painted twisted *Salix* (willow), some rolled *Aspidistra* leaves, *Bupleurum*, and large orange and pink *Gerbera*.

The butterflies were printed from the computer and I painted them with matt water-based varnish. This thickens and gives the paper a slight sheen. The next job was to cut these butterflies out by hand. (Please note that they were all held in place by magic!)

Materials Used:

Pink and orange *Gerbera*, *Salix babylonica* var. *Pekinensis* 'Tortusa' (twisted willow), *Aspidistra* leaves, *Bupleurum*.

Mother Nature's Crown

Over the years I have made many bridal headdresses for weddings but never a full crown from mixed materials and natural plant materials combined. This was an exciting project to create being another challenge and something completely new.

I was not quite sure where to start with this crown but I knew that it would need a framework to begin with onto which all the materials could be attached. After experimenting with various wires, I decided to use a mixture of rustic wire and paper-covered wire woven and twisted together to make the main framework, then painted it with POWERTEX® Bronze fabric hardener. This soaked into the paper-covered rustic wire and made it rigid. I then glued over the framework a few pieces of heavyweight lace fabric also stiffened with POWERTEX® Bronze. This gave a fantastic base to work on and it was easy to wire and glue the components on.

The next step was to wire on small pieces of corkscrew hazel stems which gives a little movement to the design and a slight wild feeling. I then wired onto the frame a few branching sprays made from an assortment of glass beads wired together with fine copper-coloured wire.

For the next stage, I had to prepare and make the rest of the components before anything else could be attached to the framework. This consisted of painting small larch cones, acorn cups, dried oak leaves and dried lichen with gold acrylic paint. I also spray-painted fresh ivy berries with gold spray paint and made a few green bullion wire flowers.

Once all the components were ready, I glued and wired everything in place with a cool-melt electric glue gun.

To complete the crown, I applied a little glitter glue to the wire flowers to make them more dominant and added a few fresh ivy leaves.

Materials Used:

OASIS® Rustic Wire, paper-covered wire, Powertex® Bronze fabric stiffener, heavyweight lace fabric, *Corylus avellana* 'Contorta' (corkscrew hazel), larch cones, acorn cups, lichen, *Hedera* (ivy) berries, ivy leaves, assorted glass beads and bullion wire.

Mushrooms and Toadstools

I find toadstools fascinating. There are hundreds of different types, all growing in a variety of shapes, colours and sizes. Many are deadly poisonous and some even look as if they have come from an alien world.

I thought that I would experiment and make a few mushrooms and toadstools using a variety of materials and colours, then arrange them to look as if they were all growing naturally in the woods.

Here I have used different craft methods, some of which I made up as I went along. The larger toadstools are carved and shaped from loft insulation foam boards, then covered with materials such as string for the stems, and dried rose petals for the tops. I used PVA glue for this as solvent-based glue will dissolve some makes of insulation board so it would be a good idea to test your glue first. When the glue had dried, I then painted them with artists' acrylic paints.

For another of the toadstools, I decorated a piece of fabric with a few glass beads stitched in place. The imitation bracket fungi I made attached to the driftwood are also carved from insulation foam and covered with thin strips of silver birch bark then painted.

The small mushrooms towards the left-hand side are crocheted knitting yarn, stiffened with gelatine then glued onto wool-covered wire stems.

The tiny mushrooms towards the right of the design are simply made from small circles of silk fabric also stiffened with gelatine, left to dry then finished with a little gold artists' acrylic paint. These are also glued onto stems of wire wrapped in coloured embroidery thread.

The three larger toadstools are decorated variously with feathers and gold paint, sisal and acrylic paint and one with papier mâché using handmade suede paper.

Materials Used:

A selection of stub wires, insulation foam, suede paper, string, silk fabric, fine knitting yarn, sisal fibre, feathers, dried rose petals, *Betula* (silver birch) bark, assorted colours artists' acrylic paint, and PVA glue.

The Woodcutter's Cache

Being a NAFAS Demonstrator and tutor, you do tend to collect lots of materials that may come in handy one day. I have quite a collection of odds and ends of bark, especially *Betula* (silver birch). These pieces I have used were part of an old fallen tree that had rotted away leaving behind the bark resembling cylinders. They make the perfect containers for a woodland theme design and would look natural and not out of place.

Our local woods have just recently been cleared of old damaged trees and the logs are neatly stacked in piles. This gave me an idea for another design.

I filled some of my silver birch bark cylinders with OASIS® Bio Floral Foam Maxlife which is perfect for soft-stemmed flowers and then placed these on the stacked logs.

The next step was to arrange the white *Hyacinthus* (hyacinth) in small groups. As they are a heavy bloom with soft stems, you will need to insert a bamboo BBQ stick into the centre of their stems for extra support. You may also need to tape part of the lower stem with florists' stem tape to hold the leaves in place.

I arranged the white tulips in a group towards the centre of the design with some upright and a few cascading downwards to give a relaxed feel to the arrangement.

I added a group of white *Ranunculus* on the right-hand side and a few dotted through. *Ranunculus* also have soft stems so you may need to make a hole in the floral foam first with a small stick or odd piece of the flower stem, then you can safely insert your *Ranunculus* without damaging the stem. (Remember that damaged flower stems do not take up water very well.)

To complete the design, I included a few groupings of green *helleborus* (Hellebore) and tucked in some fresh carpet moss.

Materials Used:

White *Hyacinthus*, white *Ranunculus*, white *Tulipa*, *helleborus*, fresh carpet moss, *Betula* (silver birch) bark and OASIS® Bio Floral Foam Maxlife.

Beauty on the Brook

How different to have a fresh flower arrangement floating in a pool of water? It would certainly create a talking point, especially at an event such as a wedding or a garden party. With a little planning, floral designs can quite easily be made to float without too much difficulty.

Here I have used a 25 cm and a 36 cm posy pad to arrange the flowers. A foam-backed posy pad (not a plastic backed) will float but will sit very low in the water so it will need some help. A good way and the easiest is to pin or glue using a cool melt glue gun some polystyrene underneath the posy pad. This will give enough buoyancy to enable your design to float just above the surface of the water. You may need to experiment with the amount of polystyrene to add. The more plant material you use the heavier your design will be therefore you will need to compensate with extra polystyrene.

I wanted to create a wild, rustic feeling, quite natural and carefree with soft textures. Firstly, I arranged the flowers in the centre of both posy pads on quite short stems in small groups. I left about 2.5 cm of bare floral foam around the outside edge of the posy pad into which I could add a mixture of fresh ivy, *Asparagus* fern, moss and natural jute string which I have unravelled to use as a fluffy texture and filler. The jute string fibres and moss also helped to conceal the edge of the posy pad. The string also acted not only as a link to both designs but also stopped them from drifting apart from each other.

Materials Used:

Asparagus fern, *heuchera* leaves, natural moss, *Hedera* (ivy) trails, *Skimmia japonica*, tulips, roses and peonies.

Chapter 5
Times Gone By

The Master's Study

Whenever you find that you have a few leftover flowers and you feel that they would look lost on display in a tiny vase, have you ever considered preserving them? It is a very simple thing to do and will not take you very long. Once these flowers are dried, you can use them in a project.

Take a few flowers at a time and fasten the end of the stems with a rubber band, then hang them upside-down somewhere warm to dry. The kitchen would be great as that is usually the warmest room in the house. Leave enough space around them for the air to circulate and this will prevent them from going mouldy and help them to dry quicker. Feel free to experiment as some flowers dry much more efficiently than others. Roses always dry well and mini sunflowers as I have used on this lamp.

I found a simple electric lamp reduced at our local DIY store and I glued in place some cinnamon sticks around the stem of the lamp with a low-melt electric glue gun. Then at the base, I glued on a selection of some of the left-over flowers that I had dried. I also repeated the flowers further up the lamp.

Once I had covered all the bare metal showing on the lamp, I gilded the dried plant material with some gold decorative wax that you apply by gently rubbing it on with your finger. (This you can find at most art shops.) To make the tassels hanging around the lampshade, I used some leftover decorative Christmas parcel cord.

Materials Used:

Assorted supermarket roses, mini sunflowers, poppy seed heads from the garden, small pine cones, *Hydrangea, Tanacetum parthenium* (feverfew), and roses made from orange peel.

The Fallen Abbey

Everyone thought that I was completely mad when I suggested we were going to decorate a wall with flowers. I like to be different and show people that nothing is impossible, especially when it comes to arranging flowers in strange places.

I was searching for an old tumbledown wall with character and a setting that would ideally portray this chapter 'Times Gone By'. I discovered this old Abbey wall locally and thought that it would be perfect. The only trouble was that it was a long way from the car park to carry heavy buckets of flowers but when you are determined enough to carry out a project you can make it possible with the help of a few friends.

Some red velvet fabric was draped over the wall and pinned up at the bottom to make it drape well. This also gave a luxurious background onto which I displayed a few old books carefully wired onto a strip of green chicken wire hidden from view. To prevent the books from falling off the wall, they were anchored down at the top with a heavy beanbag of gravel.

Next, I placed two wooden candlesticks with large church candles on the wall. These helped portray a feeling of worship and linked well with the colour of the books.

The mechanics of the flower arrangement were extremely simple and, by carefully distributing the weight of the flowers, I managed to balance the placements without the need to anchor the freshly soaked half blocks of foam in place. Space was tight so it was not possible to use dishes and bowls, so we constantly sprayed the flowers with water, so they lasted well for the photoshoot.

This design was arranged in a wild naturalistic style which I find quick to achieve. I added the flowers a variety at a time working from left to right equally distributing them throughout the design. I added a few trails of plant material to cascade down over the wall which helps to integrate the arrangement into the gap in the wall and to make it appear to sit happily.

The old ruins certainly had a spiritual feeling of 'Times Gone By' and gave us a calm and peaceful end to a wonderful late summer's day.

Materials Used:

Amaranthus 'Red Cord', *Amaranthus cruentus* 'Velvet Curtains', red *Antirrhinum*, *Cotinus*, *Hydrangea* 'Ruby Red', *Rosa* 'Black Baccara', *Rosa* 'Red Naomi', *Rosa* 'Party Trendsetter', *Zantedeschia* 'Captain Promise', *Rubus* (blackberry) stems, *Berberis*, ferns, ivy, carpet moss.

Call the Porter

Railway stations, steam and old luggage are such great subject matter when it comes to nostalgia! It brings a little piece of the past back to life and gives you that warm cosy feeling inside. The last of the steam trains were decommissioned a year or two just before I was born (Not giving away my age!) but still there are many of these beautiful engines and heritage stations open for visitors all around the country.

Early one morning I took an old hamper and a few boxes which I had decorated, along with a bucket of flowers to the Watercress Line in Hampshire. I borrowed the porter's trolley, wheeled it in front of Lord Nelson (as you do) and started to assemble a nostalgic design. It was a cold gusty day and flower arranging articles proceeded to propel themselves along the platform!

Everything had to be securely fastened with extra floral pot tape and plenty of floral fixes. The two cardboard hat boxes were decorated, one with wrapping paper I had found, and the other with dried *Aspidistra* leaves. (I try not to throw too many things away as you never know when dried items may come in handy one day!)

I chose to use these vintage colour roses as they show up well against the luggage and complement the red hat box. I picked out the red with some *Hydrangea* and a few fresh apples. I was very lucky to be able to borrow these leather gloves that match the colour of the roses so well. Once all the plant material was in place, I lightly sprayed some of the apples and flowers with an antique gold-coloured floral spray paint.

Check that your paint is suitable for use on fresh plant material by testing a sample flower or leaf first as some plant material will curl up and discolour! There are appropriate makes of spray paint suitable for this purpose on the market.

Materials Used:

Hypericum berries, *Hydrangea, Photinia* x *fraseri* 'Red Robin', *Rosa* 'Quicksand', *Hibiscus, Aspidistra* leaves and apples.

A Christmas Wedding

There is nothing like a good wedding. It is such a happy time and there is always so much activity. I jumped at the chance to be able to decorate our charming tenth century local church in Compton. The theme for this wedding was a natural, traditional feel but at the same time, grand and elegant.

My helpers and I made a huge garland from blue spruce, ivy, dried *Hydrangea* flower heads, dried *Magnolia grandiflora* leaves, pine cones, gold voile fabric and homemade baubles. We made the garlands in two halves, each on a ridged wire frame, carefully backed with a soft cloth to protect the delicate stone carved arch.

We decided to keep the overall design symmetrical in harmony with the symmetry of the building. In each corner we arranged two pedestals, also using the same materials, but including a few extras like preserved orange slices, contorted hazel and roses made from a natural plant fibre mesh.

To the left and right sides of the altar step, we created similar matching corner designs. To finish off, fresh moss and pine cones were carefully laid on the floor and altar step. (I must emphasise that great care and effort was taken to look after this ancient building. Not a drop of water was spilt!)

Materials Used:

Blue spruce, ivy, *Hydrangea, Fatsia japonica, Corylus avellana* 'Contorta' (contorted hazel), *Salix babylonica* var. *Pekinensis* 'Tortuosa' (twisted willow), *Magnolia grandiflora*, pine cones, preserved orange slices, fresh carpet moss, pine cones, voile fabric, plant fibre mesh roses and gold spray paint.

A Floral Tribute

This is something that you do not see very often, a tombstone draped with beautiful flowers. At first, I was a little worried about this idea out of respect for the graves and the wishes of the parishioners. I thought that this idea would be completely different to anything else that I had ever attempted before and another challenge to tackle.

We sought permission from our Vicar, carefully explaining what we were hoping to create in detail. The Vicar was rather intrigued with the idea and could not wait to see the finished result. Everyone was happy with the plan, so we went ahead with the design.

As this was another out-of-doors design, you can never quite predict the weather precisely so in the end, we worked under a gazebo temporarily placed over our chosen tombstone. It was meant to be a dry day but unfortunately, we had random showers. So, we worked under our shelter praying for a dry moment for the photographer to work his magic. It was also quite a surprise to some of the locals using the footpath which runs through the churchyard, so there were a few onlookers.

The top of the tombstone is covered with three floral foam designer sheets into which the flowers and foliage were arranged. In this case, all the flowers and leaves were angled towards the camera to give the design a much greater impact.

The flowers cascading down the front of the design are simply pushed into green wire mesh. I left the stems long, so their weight prevented the flower heads from dropping out. Although these flowers were in the water, it was quite a cool damp day which helped them to look their best and last long enough for the photographer. They were also sprayed with plenty of fresh water.

Towards the edges of the design, I glued dried moss onto part of the wire mesh which had been sandwiched between the duct tape. This gave a strong flexible surface onto which the moss could be glued with an electric glue gun.

To add flow and movement to the design, I added a length of cotton muslin fabric which I dyed apricot to complement the beautiful roses.

We were so lucky with the weather. As soon as we had finished the design, the sun came out and it stopped raining, so we all quickly moved the gazebo to one side so that the photograph could be taken.

Materials Used:

Rosa 'Peach Avalanche', peach carnations, peach *Chrysanthemum* blooms, assorted *Heuchera* leaves, and dried carpet moss.

Floral Teacup

I spent a lot of thinking time about this chapter, pondering over many ideas, and came up with lots of designs. I decided not to take many any further but this was one I gave a whirl! I was looking at a packet of dried skeletonised leaves I bought from a garden centre and suddenly remembered a design that I had made many years ago for a competition. It was a mask made from layers of preserved leaves glued together (Just like pâpier maché).

Here I used an old teacup and saucer and made a skeletonised leaf version of them. First, you need to lightly smear the cup and saucer with a thin coating of Vaseline, and then wrap them in cling film. This will protect the china cup and saucer and allow the glued leaf mould to slip off. Next glue on the dried leaves, layer by layer, using PVA glue. When the layers are five to six thick, leave the cup and saucer to dry. (This will take a few days.)

When your new copy of the cup and saucer is dry, take if off the china original and trim any rough edges with scissors and glue another layer of leaves over the rim of the cup and saucer. (You will have to cut around the handle of the cup to be able to release the copy, and then also patch up the hole with more leaves.) The handle can be made from wire rolled in a dried leaf glued in place.

To decorate the cup and saucer, I glued on a few pressed pansy flowers, ivy leaves and delicate leaves from an *Acer* tree. The pressed plant material was then sealed over with a couple of coats of PVA glue. Last of all, I painted a band of gold paint around the rim of the cup and the edge of the saucer.

The flowers were arranged in a tiny piece of floral foam having lined the cup so as not to get it wet.

Materials Used:

Skeletonised *Acer* leaves, *Viola* (pansy), *Hedera* (variegated ivy), *Heuchera, Alchemilla mollis, Pelargonium, Geum* (avens), rose hips, orange roses, *Dahlia* 'Moonfire', *Helenium, Ipomoea lobata.*

Homage to Grinling Gibbons

Grinling Gibbons was a seventeenth century wood carver who created the most exquisite carvings for many stately homes including the fantastic Petworth House in Sussex. They comprise panels made to imitate swags and garlands of flowers, foliage and wildlife and many other objects, all beautifully arranged. They are out of this world and must be seen!

In the world of floral design, they are a favourite and are a challenge to reconstruct, especially in natural dried plant material. So, here is my attempt at this amazing subject matter. I hope you like it!

I started with an old mirror in a wooden frame that was found in a charity shop. Before I started to construct anything, I renewed the picture wire on the reverse of the mirror as it had to take a lot of weight. As this design was quite a project, I spent a few days carefully arranging the dried plant material that I had collected over time. I would lay everything out and keep rearranging it until I was happy with the result before anything was glued in place.

I carefully stuck the heavier pieces of material with a tube of instant 'Grab adhesive' and the more delicate pieces with super glue. After all, the glue was dry and the pieces of plant material were secure, the next stage was to paint every square centimetre in brown artists' acrylic paint. This took quite some time, checking nothing was missed.

I left everything to dry for a couple of days before applying a light coating of gold wax which comes in a small tube from most good art shops. I generally apply this with my finger as you can cover the objects with accuracy and the gentlest of pressure. You need to use it sparingly as you are only highlighting the plant material, just enough to catch the light. The overall effect is an antique-looking carved wooden mirror, in the style of Grinling Gibbons!

Materials Used:

Assorted seashells, snail shells, *Fagus* (beech) mast, dried mushrooms, assorted seed pods, ribbon, wooden beads, instant 'Grab adhesive', super glue, artists' acrylic paint, and a tube of treasure gold.

Harvest Time

When I was looking for inspiration for this chapter of the book, I discovered this spot at a friend's farm and thought it would make a perfect backdrop for a design.

I was thinking of putting together something rustic but not too complicated, and to incorporate a colour to complement the green in the background. I finally chose to use red flowers in this design as it is a complementary to green which was a good starting point.

I really needed structure for a framework to arrange the flowers on and something that would give interest using various forms and shapes. I had quite a reasonable space to work with, about 2.4 metres wide. I decided to use some pieces from my collection of wooden fruit and vegetable crates and this fantastic old chair that had seen better days. After experimenting, trying out different combinations and positions of the crates, chairs and props, I came up with something suitable to arrange the flowers and foliage on.

The first step was to place all the dishes and trays of freshly soaked floral foam, and then came the fun part which was adding the plant material. As a rule, I generally add the foliage first to decide the outline shape and size of the design. Next, I placed the apples, and then added the red roses. The rest of the plant material followed.

I used quite a lot of *Asparagus* fern as it gives a great soft fluffy texture and is good for softening hard edges and breaking up dominant lines.

To finish the design, I placed some fresh moss here and there which adds to the ambience of the arrangement and is also very useful for hiding dishes of floral foam.

Materials Used:

Red roses, *Skimmia japonica*, *Hypericum* berries, *Hydrangea*, *Visnaga daucoides*, *Asparagus* fern, apples, and *Bergenia cordifolia*.

A Country Affair

This is another beautiful old church not far from me with a super porch built from amazing weathered oak timbers. I jumped at the chance to be able to decorate it for a wedding design.

You always need a starting point for any design which can be anything from a flower to a container. Here I started with this pair of characterful old rusty milk churns which I found on the internet and collected from North Wales. The tops of milk churns are quite large, and I found a plastic bowl that rested securely just inside into which I could place floral foam.

I wanted to create a country-style Autumnal theme in shades of rusty coppers, reds and oranges. I included an array of rustic textures which also link in with the country theme.

The garland above the porch is really a pair of swags that do not meet in the centre. I made them both on a ridged wire frame bent into shape and held together with strong pot tape. These were made at home with the help of a friend as there was a lot of wiring of plant material involved. The swags are made completely from dried and preserved plant material which is ideal as there is nothing to wilt in the hot sunshine.

Both milk churns are loosely arranged triangular designs very much how you would put together a large pedestal. I included some flower spikes of *Eremurus* (foxtail lily) to give height to the designs and lots of trailing plant material to make the design flow and spill down towards the ground. The focal points are spheres covered with dried lichen which I sprayed with copper and orange spray paint. I also painted the dried bracket fungi with watered-down acrylic paints to enhance their colour and make them stand out.

To complete the design, I draped some metallic copper fabric from the top of each milk churn which carries the colour of the flowers down to the ground and helps to break up the solid form of the two cylindrical containers.

Materials Used:

Garland: Dried *Triticum* (wheat), dried bracket fungi, glycerined *Fagus* (beech) leaves, *Corylus avellana* 'Contorta' (corkscrew hazel), dried thistles, *Cynara* (cardoon), roses made from hessian fabric, and mini lichen spheres.

Churns: Glycerined *Fagus* (beech) leaves, *Eremurus* (foxtail lily), spray chrysanthemums, roses, dried bracket fungi, spheres covered with dried lichen, dried and painted *Strelitzia* leaves, *Berberis, Achillea, Physocarpus opulifolius* 'Diabolo', *Clematis* trails and *Crataegus* (hawthorn) berries.

Antique Fabric

It is amazing what you can make out of fabric! The choice of colours, textures and patterns is overwhelming, and I find it so exciting when rummaging through the remnants bin at a local fabric warehouse as you never know what you may find. Also, to see all the rows of brightly coloured bolts and rolls of soft furnishings fabric is out of this world!

Here I have made a S-shaped pedestal arrangement mostly from fabric and a few other components. I selected these rich warm colours and textures to complement the rustic background.

My first task was to make a selection of flowers and foliage out of the fabric. All the roses were stitched and glued together and then mounted onto wires to enable them to be arranged in a dry foam block. I made the foliage from a heavyweight brocade fabric that had been stiffened, and left to dry then the leaf patterns were cut out and glued onto wire stems.

The dry floral foam was secured to the top of a metal stand disguised with some more fabric and some voile. To give an aged, luxurious appearance to the completed design, I painted on a little bright gold acrylic paint which also highlights the different shapes of the flowers and gives added interest. The tassels are made from polystyrene spheres covered in red velvet remnants and gold curtain trimmings.

To complete the design, I scattered fallen autumn leaves beneath the arrangement.

Materials Used:

Assorted curtain and upholstery fabric, knitting yarn and cord.

Miss Havisham's Wedding

I have always wanted to recreate the scene from Miss Havisham's wedding table ever since I first saw it in an old black and white movie many years ago on television. Miss Havisham is one of Charles Dickens's characters in his famous book *Great Expectations*. In the story, Miss Havisham was jilted on her wedding day and remained in her wedding dress for the rest of her life and her laid table was left untouched. It certainly is an exciting title that you could get your teeth into. So, I did!

This was not an easy subject matter to put together as it is made up of many different components which took a long time to produce. I possibly spent about two months working on this project.

It all began by researching Victorian table settings and the obvious choice was Mrs Beeton's famous book *Household Management*. This helped get the ball rolling by giving me some ideas of what to start collecting and assembling, something rather formal but faded and covered with cobwebs.

I then spent a while collecting flowers for drying, fabrics, paints and various other materials. Grey cotton sheets were painted with spray paints and splats of acrylic paint to form the floor and a false wall which helped to create the scene.

On the centre of the table, I set a large silver candelabrum and four smaller ones at each corner. It was on these that I attached four dried flower garlands. These have also been coloured and antiqued with acrylic paint and spray paints. In front of the large central candelabrum, I placed the wedding cake which I made from polystyrene iced with white tile grout and 'No More Nails' adhesive. I then coloured with artists' acrylic paints and decorated with paper and homemade flowers made from various materials including ribbon.

Above the table, I hung a birdcage with a homemade blackbird inside. I thought this would add some charm and character to the design and would also help to create interest towards the top. As I could not find a suitable birdcage, I made this myself from Rustic Wire and hung cobwebs around. The blackbird was constructed from textiles, which was quite a challenge!

A bowl of fruit in a raised dish ensured that not everything on the table was at the same level. Most of the fruit is plastic which I had spray painted in a complementary colour. I included a few dried artichokes painted to match the main colour theme of the table setting. They are a wonderful form and add interest.

Once all the dried flower garlands and hand-tied sprays were in place, I then added a few fresh flowers and foliage which brought some life to the whole design as well as more colour.

One of the most enjoyable parts of creating this design was stretching out and placing the cobwebs then covering the whole set with a dusting of dry cement. The cobwebs are made from a bag of polyester fibre.

Materials Used:

Fresh Plant Material: *Rosa* 'Ocean Song', *Rosa* 'Memory Lane', mauve *Hydrangea, Heuchera* leaves, *Hedera* (ivy) trails.

Dried Plant Material: *Achillea*, raffia, *Morus* (mulberry) bark, *Gypsophila*, roses, skeletonised leaves, and sisal.

Left Luggage

They say you can tell a person's character by their luggage. I wonder what sort of person left this in the railway waiting room, maybe we will never know but it's fun trying to guess!

While working on some ideas for a railway theme design, I kept looking at a bunch of dried flowers that were sitting on the shelf in my shed. I thought that all the flower heads packed together looked just like a section of an old-fashioned carpet. Then another idea was hatched, an old Edwardian carpet bag! Just the sort of thing that would be left on the hard-wooden bench of a waiting room.

Underneath is a cardboard box, cut and glued together to form the basic shape of a bag. I lined the inside of the bag with preserved oak leaves, held in place with hot melt glue. I then glued on the flowers that I had dried to form a simple pattern, complemented by a few swirls of thick twine painted gold.

The colours that I used are all soft muted tones that give the overall effect of an aged, well-travelled carpet bag. The gaps were then filled in with dried moss also glued on with my trusty glue gun. The handles are made from palm bark stitched together with a gold cord.

I was really pleased with this design and thought it turned out well and, as a bonus, it also cost almost next to nothing to make!

Materials Used:

Dried carpet moss, assorted dried roses, dried peonies, dried *Hydrangea* and palm bark.

A Final Word

Sadly, now we have come to the end of my book but not the end of the journey. The journey so far has certainly been the trip of a lifetime. It has been so amazing I could write a book about the production of this book. It all started so innocently at the beginning seven years ago when one day I was showing a friend some pictures of my work and she suggested that I should write a book, so I did and here it is. I thought I could write this book in six months; I had no idea of the work involved.

The British weather has been one of the greatest challenges as many of my floral designs have been made and photographed out of doors. One memorable event was a design on the beach. We arrived in my car loaded with flowers, photography equipment, a photographer, and a floral assistant. After religiously following the weather forecast, we decided it would be all systems go. We arrived at the beach, parked, and could not open the car door. The wind was so strong the door kept being blown shut! Anyway, we eventually found a sheltered spot and managed to make a great design.

I also cannot believe how many cups of tea and slices of cake we got through after each photoshoot! We always treated ourselves after all the hard work carrying buckets of flowers and photography equipment over sand dunes, through hedges and across car parks.

I discovered a huge amount of arranging for the camera, how to choose the most perfect locations, the best time of the day to photograph and what angles work best. When I first began this book, I would try to do too much in one day and wondered why designs looked rushed and incomplete. You really cannot rush designs, you must take your time, and include lots of tea breaks (wherever possible) then the finished results will be perfect.

Had I known how much work was involved, I possibly would not have jumped in at the deep end. It has also been so incredibly frustrating at times as a few designs had to be remade two or three times as they just would not photograph well. Sometimes the colours would not work, sometimes the shape of the design looked completely strange, sometimes the mechanics failed, and the design fell over but still, you must keep on going.

After seven years of experimentation with lots of ideas, I feel that you never really stop learning. There is always something new and exciting to try out. I am certainly very pleased with the results of the many hours of very hard work put into this project and especially by my friends. I do hope that you are equally impressed with my book and I thank you for letting me share my work with you. Maybe I have inspired you to have a go, experiment and create your own designs and ideas. Perhaps you might visit your local NAFAS (National Association of Flower Arrangement Societies nafas.org.uk) flower club, group, society, and visit a flower festival or even join a craft group.

Possibilities are endless!

Graham King.

www.ingramcontent.com/pod-product-compliance
Lightning Source LLC
Chambersburg PA
CBHW040108180526
45172CB00009B/1267